Back to Newton

Back to Newton

Everything should be made
as simple as possible,
but not simpler.

Albert Einstein

ex nihilo nihil fit

Jan Slowak

Back to Newton

Earlier books

1. Bye-Bye Big Bang, Episod/Episode 1
2. Bye-Bye Big Bang, Episod/Episode 2
3. Bye-Bye Big Bang, Episod/Episode 3
4. Redshift factor, Absolute redshift, Galaxies red / blue distribution
5. Sawing of my article about the Big Bang
6. Big Bang - Questions to physicists and cosmologists
7. Einstein's theory of special relativity - mathematical and physical mistakes! (swedish)
8. Back to Newton (swedish)

Copyright © Jan Slowak 2017
Publisher: BoD - Books on Demand, Stockholm, Sweden
Printed by: BoD - Books on Demand, Norderstedt, Germany
ISBN: 978-91-7699-413-9

Back to Newton

*For
Science*

Content

1) Bibliography 7
2) Prolog 9
3) Historical rewiev 13
4) Everything is relative 15
5) Events in the coordinate system 17
6) Light 19
7) Registration, calculation and transformation of coordinates 21
8) Analysis of time dilation 29
9) Derivation of the Lorentz Transformations 1 39
10) Derivation of the Lorentz Transformations 2 43
11) Derivation of the Lorentz Transformations 3 45
12) Derivation of the Lorentz Transformations 4 49
13) Derivation of the Lorentz Transformations 5 53
14) Derivation of the Lorentz Transformations 6 57
15) Derivation of the Lorentz Transformations 7 61
16) Michelson-Morley experiment 63
17) Epilogue 69
18) One of my articles... 73

Bibliography

[1] Modern Physics; Sixth edition; Paul A. Tipler, Ralph A. Llewellyn; Chapter 1; Relativity I; 2012

[2] University Physics with Modern physics; Thirteen Edition; Young Freedman; Chapter 37; Relativity; 2012

[3] Den speciella och den allmänna relativitetsteorin; Albert Einstein; Första delen; Om den speciella relativitetsteorin; 2006; (swedish)

[4] Einsteins relativitetsteori – en kritisk analys ...; Ove Tedenstig; 2015; (swedish)

[5] Den moderna fysikens grunder ...; Krister Renard; Kapitel 2; Speciell relativitetsteori; 1995; (swedish)

[6] Concepts of Modern Physics; Sixth edition; Arthur Beiser; Chapter 1; Relativity; 2003

[7] Modern Physics; Second edition; Randy Harris; Chapter 2; Special Relativity; 2008

[8] Knowing, The Nature of Physical law, Michael Munowitz, 2005

[9] Illustrerad vetenskap, Nr 16/2014; (swedish)

[10] Calculus - A Complete Course; Robert A. Adams; Sixth Edition;

[11] Nádherná teorie – Sto let obecné teorie relativity; Pedro G. Ferreira; (czech)

[12] Six Ideas That Shaped Physics; Thomas A. Moore; 2003

[13] Calculating the cosmos; Ian Stewart; 2016
...

Prolog

When I started researching for real the theory of special relativity, I went thoroughly through most of the books I found in the university library. I mean that I read everything that concerned the theory of special relativity. Then it was the articles on the web. Some of those told that all scientists do not agree on this theory.

Can one say research for real if one have a job that has nothing in common with what you want to do research on? In any case, all my spare time went for this purpose. One may wonder why I was doing so. For me it was an old desire. The first time I got in touch with the theory of relativity was in high school. And I could *not* accept it. Not time dilation, not the twins paradox, not the length contraction.

The time went by. I studied mathematics and computer science at the university. And since then, I have worked as a software developer, programmer.

So why should *I* do research on the theory of special relativity? Why so late? Well, I was always overwhelmed by the science. I was overwhelmed every time I was reading about scientists who came up with

new findings and explanations how things work in different fields: anthropology, genetics, astronomy, cosmology.
No, not cosmology, not the expanding of the universe, not the Big Bang, not the dark matter.

But why could I accept most of the new ideas except those in cosmology?
Because my motto was that we learned in school:

ex nihilo nihil fit

I started my research in cosmology and the theory of special relativity sometime in 2014. It was the analysis of data from the database NED Redshift-Independent Distances from
http://ned.ipac.caltech.edu/Library/Distances/

The result of this research I published in my book *Redshift factor, Absolute redshift, Galaxies red/blue distribution*. And the result was astounding, in my opinion:

Population	zf	Num obj	Num red z	% red z	Num blue z	% blue z
NED-D	0.000239	26,790	13,018	48.6	13,772	51.4

We see here that the distribution of the objects' redshift and blue shift is about 50/50!
The Big Bang theory says that *most* cosmic object has a redshift, except for some in our neighborhood that may have blue shift. My research shows that there is no argument for the expansion of the universe!

I sent my book to some researchers. The book was sawed! One could say that the result was based on my own interpretation of the data.

Therefore, I decided to go to the source of the problem. Big Bang theory was based on Einstein's theory of relativity.

The result of this research, I published in my book *Einstein's theory of special relativity - mathematical and physical mistakes! (swedish)*

But today, anyone can publish a book. The question is whether you get recognition for your ideas and your research. It is a difficult task! It's like fighting against the "windmills"!
I wrote a few articles based on my book and sent them to some magazines. I sent question to a number of institutions if I could present my research to some researchers. I got no answer or the answer was negative!

In this book I intend to summarize my research on the theory of special relativity. I will come up with evidence that the theory of special relativity is wrong fundamentally, in its entirety!

Historical review

We present some scientists that somehow were mentioned when talking about the theory of special relativity.

Galileo Galilei, 1564-1642
Galilean transformation, $x' = x - vt$

Isaac Newton, 1642-1727
Time is universal and the same everywhere
The room is the same in all places, and the same in all directions
The room is homogeneous and isotropic
Absolute motion

James Clerk Maxwell, 1831-1879
Maxwell equations
The speed of light, $c = 1/(\mu_0 \varepsilon_0)^{1/2}$

Albert A. Michelson, 1852-1931
Michelson-Morly experiment, 1887
The light-bearing ether

Hendrik A. Lorentz, 1853-1928
Lorentz transformation, $x' = (x - vt)\gamma$, $t' = (t - vx/c^2)\gamma$
Lorentz factor, $\gamma = 1/(1 - v^2/c^2)^{1/2}$

Albert Einstein, 1879-1955
The theory of special relativity

In this book I show that the set-up of the Michelson-Morley experiment of 1887 was incorrect, and therefore why the conclusion about ether was wrong.

The negative result of the Michelson-Morley experiment was followed by the Lorentz transformations and the special theory of relativity.

I show in seven different variants that the derivation of the Lorentz transformation is incorrect.
This transformation is the foundation of Einstein's theory of special relativity.

My conclusion is that the theory of special relativity is incorrect in its entirety!

Everything is relative

When talking about the relativity, it is about how one observer perceives things with the help of the information the observers receive with their senses: touch, hearing, sight. To say that it's cold out there can mean for an observer to tremble from the cold, but another observer would say that it is quite comfortable out there. But when we use a thermometer and measure the temperature to -5 degrees then it is -5 degrees. It is a physical measurement. A thermometer does not perceive temperature, it measures the temperature! What the two observers than even are saying about how cold it is out there they have to agree that it is -5 degrees, no more comments!

To perceive events and measuring their coordinates are two different things. Therefore, it feels weird every time I read about the thought experiment with two observers, one of which is still standing on the platform and the other sitting on the train that is moving at a constant speed toward the platform.
In this book, I will describe these thought experiments to define what happens physically, and not how someone observer perceives it one or the other.
The theory of special relativity treats including coordinate systems, concurrency, event, time, place,

Lorentz transformations, reference systems, observer, time dilation, thought experiments, and other concepts.

Events in the coordinate system

An event in spacetime is specified by 4 coordinates. We denote an event with the letter E (from Event). Such an event can be denoted as follows:

$$E = (x, y, z, t)$$

To simplify things, we only consider the events taking place on the x-axis. Then $y = 0$, $z = 0$ and then we denote the event only by

$$E = (x, t).$$

In these experiments, we will use the material objects that can transmit a light signal and that can record an incoming light signal. Such object on the x-axis is a coordinate system. We denote them by S, S_1, S_2, S' and so on.
We say that they are **material objects** to distinguish them from the light signals which are **wave phenomena**.
The coordinate systems used in our experiments can be stationary to each other or move relative to each other at a constant velocity, $v > 0$.

The information between these systems is mediated by light signals, moving at the speed of light c. We approximates c to 300 000 km / s.

Light

Light, like other electromagnetic radiations, is a wave phenomenon wich propagates in space and time. Light moves regardless of how the source or the observer moves.

But also the direction that the light signal is moving in is independent of how the source or the observer moves.

It doesn't matter if the light source moves or rotates, in the moment the signal leaves the light source, the light signal moves with the same speed and the same direction.

We illustrate how the speed and direction of the light signal are independent of how the light source moves, see Fig. 1.

We consider S_1 that transmits a light signal every microsecond, while the source S_1 is turning with an arcsecond. In a microsecond the light signal travels a distance of *0.3* km. At a distance of *97,200* km there is the S_2. When S_1 is facing S_2, first light signal is transmitted. After *324,000* microseconds (90x60x60) the light signal reaches S_2, and S_1 is turned *90* degrees left/right. And it is only the first signal that reaches S_2!

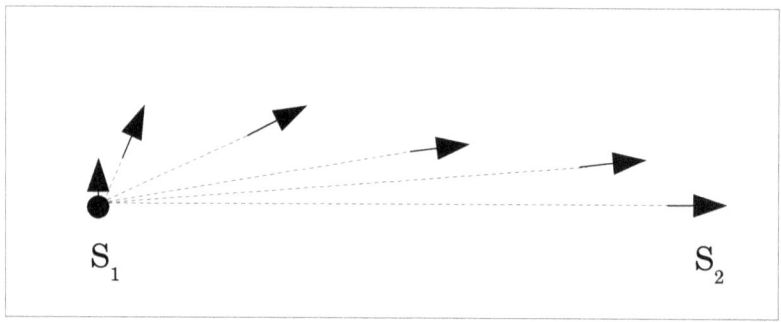

Fig. 1

Registration, calculation and transformation of coordinates

The theory of special relativity treats two coordinate systems that move relative to each other with constant speed, $v > 0$. The theory of special relativity tell us how to calculate the coordinates of an event in one system using coordinates from the other. Such calculation is called transformation.

Let's look first at a coordinate system and an event, Fig. 2. An event E occurs in the coordinate system S_1 at time t. S_1 get information about the event by registering the light signal from it. **Full details of the event we have only if we know the event's *x*-coordinate, the distance from S_1 to E.**

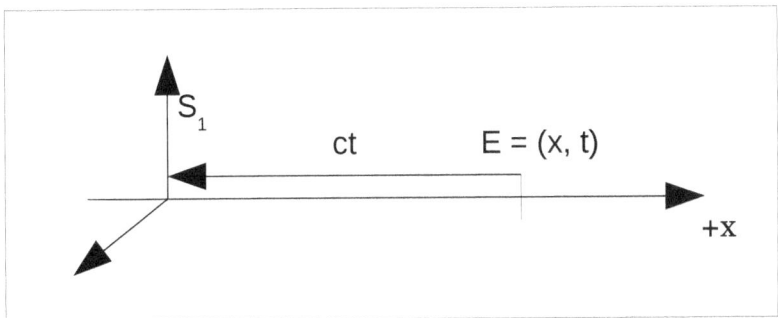

Fig. 2

Then $t = x/c$, and we can designate the event with

$$E = (x, x/c)$$

Now we look at two coordinate systems, S_1 and S_2, standing relative to each other, and an event E. See Fig. 3. Distance between S_1 and S_2 is d.

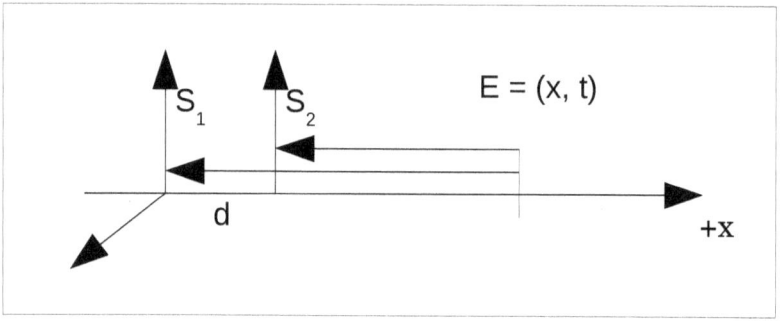

Fig. 3

How does the event $E = (x, t)$ look like when it is registered in S_1 and S_2?

$$E_1 = (x_1, t_1) = (x, x/c)$$
$$E_2 = (x_2, t_2) = (x-d, (x-d)/c)$$

If we know the distance between S_1 and S_2, we can calculate the coordinates of the event in one of the coordinates of the event from the other. For example:

$x_2 = x_1 - d$ och $t_2 = t_1 - d/c$

How will it be when S_2 moves to the right with a constant velocity $v > 0$? See Fig. 4.
The experiment starts at $t = 0$. S_1 and S_2 at that time are in the same point, $x_1 = x_2 = 0$.

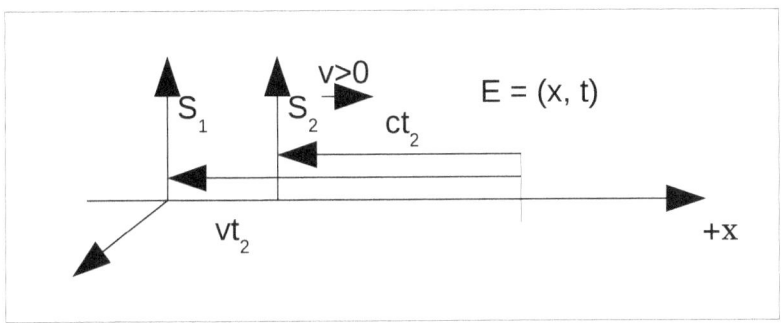

Fig. 4

The only difference between Fig. 3 and Fig. 4 is that instead of distance d, we have distance vt_2.
We have $E_1 = (x_1, t_1) = (x, x/c)$ and using x_1 and t_1 we calculate $E_2 = (x_2, t_2)$. Here we use the fact that the time it takes for the light signal to reach S_2 is the same as the time S_2 needs to cover the distance from the point (0, 0) to the point where it meets the light signal. Then we have

$$x = ct_2 + vt_2 \rightarrow t_2 = x/(c+v).$$
$$E_2 = (x_2, t_2) = (x_1 c/(c+v), \ t_1 c/(c+v))$$

So both x- and t-coordinate are calculated by the same factor **c/(c+v)**.
In the example above, we have placed the event E in **front** of the S_1/S_2, if you think about the direction in which S_2 is moving.

Now we place the event **behind** S_1/S_2, see Fig. 5.
In this thought experiment the x-coordinates x, x_1 and x_2 are negative. t-coordinates t, t_1, t_2 are positive, always.

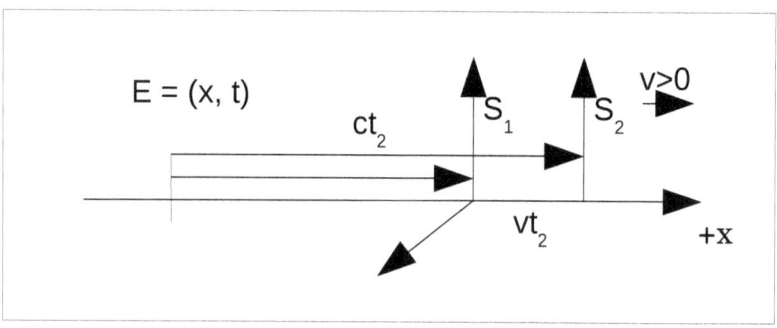

Fig. 5

We have $E_1 = (x_1, t_1) = (x, -x/c)$ and using x_1 and t_1 we calculate $E_2 = (x_2, t_2)$. Here we use the fact that the time it takes for the light signal to reach S_2 is the same as the time S_2 needs to cover the distance from the point (0, 0) to the point where the light signal reaches S_2.

This time we have

$$-x = ct_2 - vt_2 \rightarrow t_2 = -x/(c-v)$$

$$E_2 = (x_2, t_2) = (x_1 c/(c-v),\ t_1 c/(c-v))$$

So both *x*- and *t*-coordinate are calculated using the same **factor c/(c-v)**.

We see that the transformation factor is not the same in the two cases, Fig. 4 and Fig. 5, the transformation is **dependent** of where the event happens!

We summarize these two thought experiments to show that the transformation factor between the two inertial reference system is not the same across the +*x*-axis.

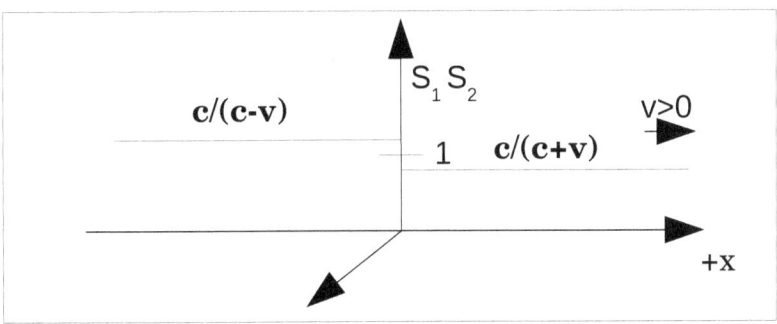

Fig. 6

Back to Newton

We do not need the Lorentz transformations, we will manage the transformation of coordinates between two inertial referens systems by classical physics!

Summary of the coordinates in Fig. 3, 4, and 5.

$v = 0$	$v > 0$

The event E is in "front" of S_1/S_2

$x_1 = x$	$x_1 = x$
$t_1 = x/c$	$t_1 = x/c$
$x_2 = x-d$	$x_2 = x-vt_2$
$t_2 = (x-d)/c$	$\mathbf{t_2 = (x-vt_2)/c = t_1-vt_2/c}$
$x_2 = x_1-d$	$x_2 = x_1 c/(c+v)$
$\mathbf{t_2 = t_1-d/c}$	$t_2 = t_1 c/(c+v)$

The event E is "behind" of S_1/S_2

$x_1 = x$	$x_1 = x$
$t_1 = -x/c$	$t_1 = -x/c$
$x_2 = (x-d)$	$x_2 = (x-vt_2)$
$t_2 = -(x-d)/c$	$\mathbf{t_2 = -(x-vt_2)/c = t_1+vt_2/c}$
$x_2 = x_1-d$	$x_2 = x_1 c/(c-v)$
$\mathbf{t_2 = t_1+d/c}$	$t_2 = t_1 c/(c-v)$

One of the main conclusions of this analysis is that the transformations are **not linear** over the hole +x-axis. We also see that if we put $v = 0$ in the transformations of x_2, t_2 (second column in the summary), we get not formulas for the "standing" system (when S_1 and S_2 are standing against each other; first column in the summary).

The fact that the transformations are not linear over whole +x-axis is one of the reasons that I continued to investigate further.

The above two experiments, Figs. 4 and Figure 5, and its conclusions should serve as a eye-opener for every scientist who works with the theory of special relativity.

This led me to conclude that the theory of special relativity is inaccurate, and that it therefore is incorrect in its entirety!

In the following chapters of this book I make the analysis of various aspects of the theory of special relativity. The analysis shows errors in the interpretation of the propagation of light, in how the Lorentz transformation was derived.
It is about fundamental physics and math!

Back to Newton

Analysis of time dilation

In some of the literature [1], [6], [8], which deals with the theory of special relativity, time dilation is explained as follows, and one uses the same thought experiment to derive the Lorentz factor.

In these thought experiments one uses spaceship in which a light beam starts on the floor, is reflected in the ceiling and comes back to the floor. We illustrate the two cases.

The first case is when the spacecraft is stationary, see Fig. 7. Distance from the floor to the ceiling is L. Then the time that the light need to cover the distance floor-ceiling-floor is

$t_0 = 2L/c$

Fig. 7

The second case is when the spacecraft is moving with constant speed $v > 0$ to the right, Fig. 8.

We consider triangle with given sides and calculate from there:

$$t = 2L/(c^2-v^2)^{1/2}$$

One replaces *2L* with $t_0 c$ and we get

$$t = t_0\, c/(c^2-v^2)^{1/2} = t_0 \gamma \text{ where } \gamma \text{ is Lorentz factorn.}$$

This is what the theory of special relativity say.

Here, I think that Fig. 8 is the most absurd, dissolve from reality, explanation of a physical phenomenon I've seen so far!

Why compare Fig. 7 with Fig. 8?
Why manipulate the experiment?
If in Fig. 7, the light signal goes straight up so should it go straight up also in Fig. 8 too! Then you can compare them!

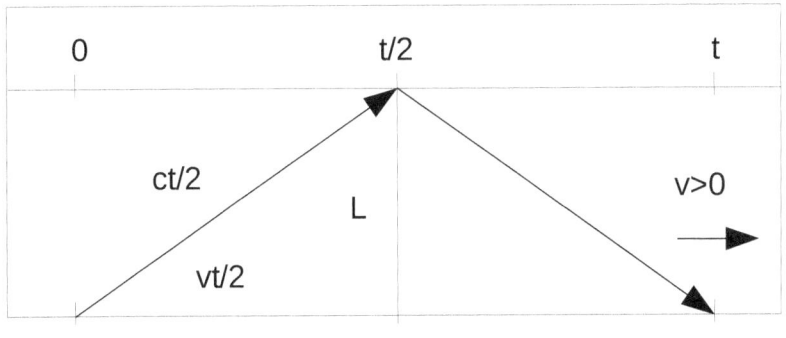

Fig. 8

***My explanation:
A light beam moves at a constant speed c and
with the same direction regardless of how the
light source moves.***

Imagine a stationary platform in the vacuum of space.
A light signal leaves the platform and will
move with the same direction, see Fig. 9.

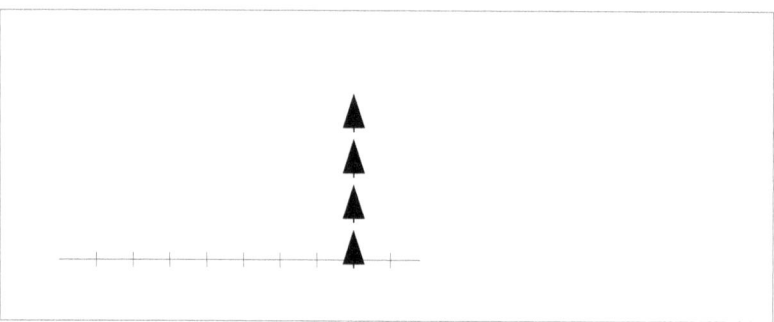

Fig. 9

Consider now the same platform in vacuum, in space, Fig. 10, moving at speed $v > 0$ to the right. A light beam leaves the platform and will move in the same direction.

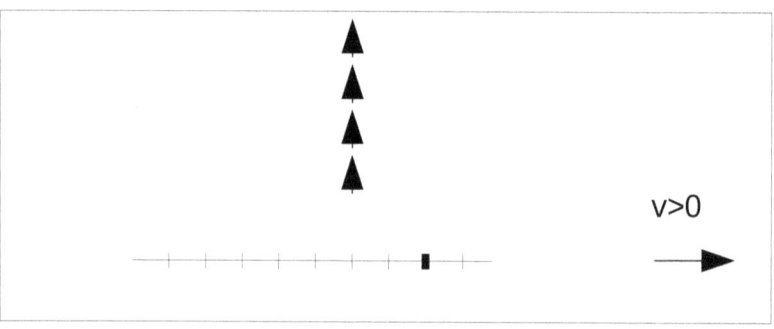

Fig. 10

We illustrate the reasoning that a light signal leaving the floor, reflecting in the ceiling and hits the floor again, is moving in the same direction.
We will, in the same image, Fig. 11, show more intermediate positions so that in a simple way see how the light signal and the "spaceship" moves.

We have a "spaceship" moving with constant velocity $v = 30\ km/s$ to the right. We consider a light signal is leaving the floor, reflects itself in the ceiling and reaches the floor again. During this time the ship moves with a distance $d = 2x$.

Back to Newton

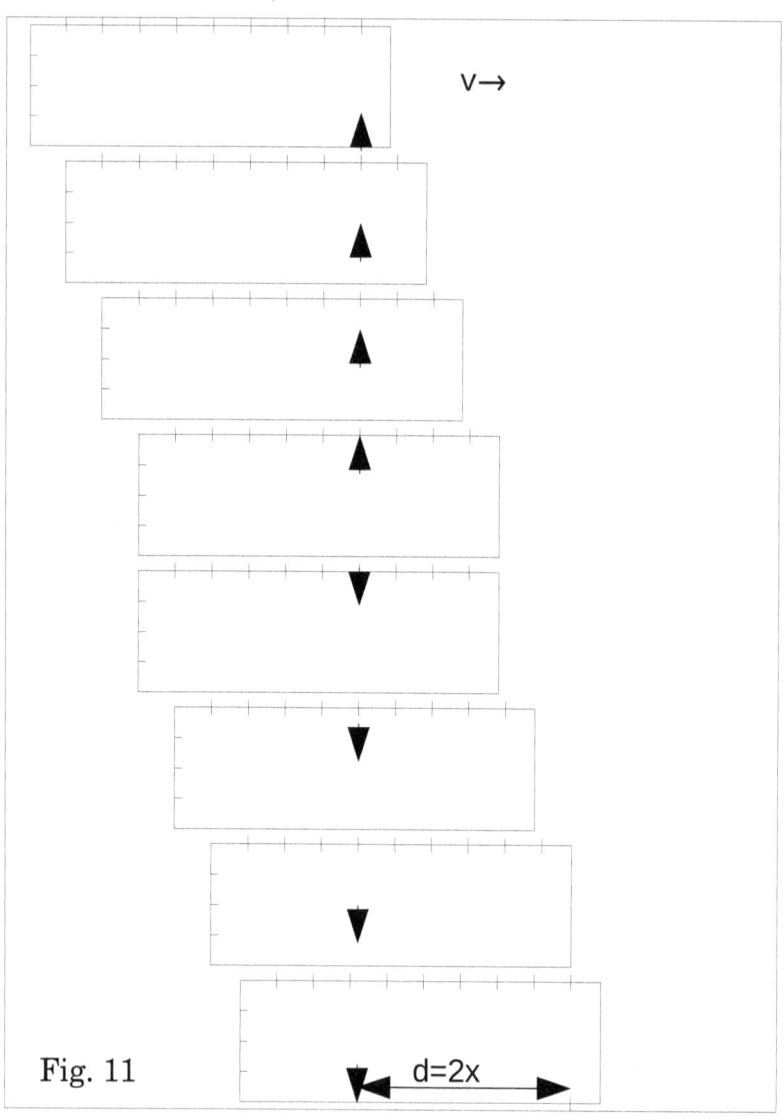

Fig. 11

Carefully consider the picture! A light signal starts from the floor, reflects itself in the ceiling and end up in another point on the floor, **behind** the point from which it started, if you think about the direction of movement.

The light propagates not in zigzag.

Distance between two points only tell us how far the ship moved in the same time as the light signal traveled distance $2L$. We denote this distance $d = 2x$.

We summarize this: The time during which the light signal had passed distance $2L$ is the same as the spaceship needs to cover the distance $2x$.

$t = 2L/c = 2x/v \rightarrow x = Lv/c$

Example: $L = 10$ m, $v = 30$ km/s, $c = 300\ 000$ km/s

$x = 10*30/300000$ m $= 1/1000$ m $= 1$ mm

This means that you could build a device that would measure the speed of the Earth in space, around the Sun, around the galactic center.

$v = xc/L$

We can build a device that can measure the absolute motion! Hence the book's title!

This device would work as an electromagnetic gyroscope, a light gyroscope.

Note that the time which the light signal travels distance *2L* is the same, either the system is at rest or is moving at a constant speed *v > 0*!

Then we have no time dilation!

Below I present six different calculations showing that the derivation of Lorentz transformations is incorrect.

When we study the physical phenomena, we always make a mathematical model of them. In such a model are built the current physical laws held together by mathematical tools. If the description of the physical phenomenon is correct, the mathematical model has no errors!

The theory of special relativity deals with the relationship between two inertial reference systems, S and S', moving towards each other with constant velocity $v > 0$. Every event in the reference system is determined by four coordinates, three for room and one for time. To determine the event's coordinates in one of the reference system with the help of the event's coordinates in the second one uses Lorentz transformations:

$E = (x, y, z, t)$, an event in S
$E' = (x', y', z', t')$, an event in S'

To facilitate understanding of the calculations one usually puts $y = y'$ and $z = z'$. Then the Lorentz transformations become:

$x' = (x - vt)\gamma$ (LT$_1$)
$t' = (t - vx/c^2)\gamma$ (LT$_2$)

where $\gamma = 1/(1 - v^2/c^2)^{1/2}$ is called the Lorentz factor.

Back to Newton

Back to Newton

Derivation of the Lorentz transformations: example 1

The theory of special relativity makes use of Lorentz transformations to calculate the coordinates of events in a reference system using coordinates in a different reference system moving towards one another at a constant speed, $v > 0$.

We follow the reasoning and calculations from *[7], page 14-15*. Here, one use the following:

(2-3) $x' = u't'$ and $x = ut$
(2-4) $x' = Ax+Bt$, $t' = Cx+Dt$

They say that between (x, t) and (x', t') must be a **linear** transformation. This in turn means that A, B, C, D are constants.
To determine the above four constants, one uses three special cases.

(S <--> S_1, S'<--> S_2)

c1)
The object in which the event E occurs is in origin of S_2.
$E_2 = (x_2, t_2) = (0, t)$

c2)
The object in which the event E occurs is in origin of S_1.
$$E_1 = (x_1, t_1) = (0, t)$$
c3)
One equates the **object** in which the event E occurs with a **beam of light**.

We follow the calculations:

c1) $x' = 0$, $x = vt$
One replaces those in (2-4) and get:

$$0 = Avt + Bt \text{ and } t' = Cvt + Dt \rightarrow$$
$$\mathbf{B = -Av \text{ och } t' = Cvt + Dt}$$

c2) $x = 0$, $x' = -vt'$
One replaces those in (2-4) and get:

$$-vt' = Bt \text{ and } t' = Dt$$

One divides these two equations and gets

$$\mathbf{B = -Dv \rightarrow D = A}$$

My addition:
But t' = Cvt+Dt from c1 and t' = Dt from c2 →
    Cvt+Dt = Dt → Cvt = 0 → **C = 0**

Then it becomes
(2-4) x' = Ax-Avt and t' = At or
    x' = A(x-vt) and t' = At

Now we use
c3) x' = ct' och x = ct
We replace these in x' = A(x-vt) and we get

    ct' = A(ct-vt) and t' = At

From here, we get:
    ct = ct-vt → vt = 0

If t = 0 then S_1, S_2 are at the same point, nothing moves.
    → **v = 0**

You get contradiction with start conditions.

Back to Newton

Derivation of the Lorentz transformations: example 2

This derivation is in *[7], page 14-15*.

The derivation of the Lorentz transformations are made with the assumption that these transformations must be linear:

x' = Ax + Bt
y' = Cx + Dt, where A, B, C and D are constants.

To solve this equation system one uses three special cases:

c1) x' = 0, x = vt
c2) x = 0, x' = -vt'
c3) x = ct and x' = ct', where c is the speed of light

and finally one comes to the Lorentz transformations

$$x' = (x - vt)\gamma \qquad (LT_1)$$
$$t' = (t - vx/c^2)\gamma \qquad (LT_2)$$

where $\gamma = 1/(1 - v^2/c^2)^{1/2}$ is called the Lorentz factor.

But if Lorentz transformations LT1, LT2 where produced using c1, c2 and c3, these three special cases must verify Lorentz transformations LT1, LT2 without mathematical contradiction.

My mathematical proof:
From c1 and LT_1
$\rightarrow 0 = (vt - vt)\gamma \rightarrow 0 = 0$, OK
From c1 and LT_2
$\rightarrow t' = (t - v(vt)/c^2)\gamma \rightarrow t' = t(1-v^2/c^2)\gamma$
From c2 and LT_1
$\rightarrow -vt' = (0-vt)\gamma \rightarrow -vt' = -vt\gamma \rightarrow t' = t\gamma$
From c2 and LT_2
$\rightarrow t' = (t-v0/c^2)\gamma \rightarrow t' = t\gamma$, same results

But the result from c1 and LT_2 is $t' = t(1-v^2/c^2)\gamma$
and the result from c2 and LT_2 is $t' = t\gamma$

$\rightarrow 1-v^2/c^2 = 1 \rightarrow \mathbf{v = 0}$

This result, *v = 0*, is in contradiction with the theory's assumption that the two reference systems move towards each other with constant velocity *v > 0*!

This shows that the theory of special relativity contains inaccuracies.

Derivation of the Lorentz transformations: example 3

Below we follow *[3], page 125; Appendix;*
A simple derivation of the Lorentz transformation

In his presentation of the theory of special relativity, Einstein finally comes to Lorentz transformations:

$$x' = (x - vt)\gamma \qquad (LT_1)$$
$$t' = (t - vx/c^2)\gamma \qquad (LT_2)$$

where $\gamma = 1/(1 - v^2/c^2)^{1/2}$ is called the Lorentz factor.

I will quote Einstein and analyze what he says.

Einstein:
"A light signal which is running along the positive *x*-axis moves corresponding the equation

$$x = ct \text{ or } x\text{-}ct = 0 \text{ "} \qquad (1)$$

The expression "along the positive *x*-axis" means that the equations (1) is valid for $x \geq 0$.
Similar applies to the second coordinate system.

$$x' = ct' \text{ or } x'\text{-}ct' = 0 \qquad (2)$$

Equations (2) applies to $x' \geq 0$.

Einstein:
"The points in space-time (events) that satisfy (1) must also satisfy (2). This is obviously the case if we generally have the relationship

$$(x'\text{-}ct') = \lambda(x\text{-}ct) \qquad (3)$$

where λ is a constant. For according to (3) becomes *x'-ct' equal to zero if x-ct is equal to zero*."

Einstein:
"An analog reflection of a light beam propagating along the negative x-axis gives the condition

$$(x'+ct') = \mu(x+ct)" \qquad (4)$$

This section applies to $x \leq 0$ and $x' \leq 0$.

Equation (3) applies to **$x \geq 0$** and to **$x' \geq 0$**.
Equation (4) applies to **$x \leq 0$** and to **$x' \leq 0$**.

Einstein:
"If one now adds and subtracts the equations (3) and (4) one obtains:

    x' = ax-bct
    ct' = act-bx"

and so on...
Furthermore, we needn't analyze Einstein's derivation of the Lorentz transformations.

Einstein makes here a fundamental mathematical error: one adds and subtracts equations that apply in completely different application areas.

I invoke *[10], page 32:*
"If *f* and *g* are functions, then for every *x* belonging application areas for both *f* and *g*, we define functions *f* + *g* ..."

We can do operations on functions only in their common application areas.

The above areas, equations (3) and (4), has a single point in common:

    x = 0, x' = 0.

But then, from (1) → t = 0, and from (2) → t' = 0, and then we have the trivial example when both coordinate systems are in the same point!

Then we cannot talk about the two reference systems moving with constant velocity $v > 0$ against each other! Then, we needn't transformations to go from one reference system to the other! They are identical. Then we need no theory that deals with the relationship between these two coordinate systems!

Derivation of the Lorentz transformations: example 4

Below we follow *[3], page 125; Appendix;*
A simple derivation of the Lorentz transformation

In his presentation of the theory of special relativity, Einstein finally cames to Lorentz transformations:

$$x' = (x - vt)\gamma \qquad (LT_1)$$
$$t' = (t - vx/c^2)\gamma \qquad (LT_2)$$

where $\gamma = 1/(1 - v^2/c^2)^{1/2}$ is called the Lorentz factor.

I quote Einstein and analyzes what he says:

Einstein:
"A light signal which is running along the positive x-axis moves corresponding to the equation

$$x = ct \text{ or } x-ct = 0 \qquad (1)$$

Similar applies to the second coordinate system.

$$x' = ct' \text{ or } x'-ct' = 0 \qquad (2)$$

"The points in space-time (events) that satisfy (1) must

also satisfy (2). This is obviously the case if we generally have the relationship

$$(x'-ct') = \lambda(x-ct) \qquad (3)$$

where λ is a constant. For according to (3) becomes *x'-ct'* equal to zero if *x-ct* is equal to zero."

Here you must specify that

$$\lambda \mathrel{!=} 0 \qquad (3.1)$$

Because if $\lambda = 0$, one can **not** say that *x'-ct'* is equal to zero if *x-ct* is equal to zero."
Because if $\lambda = 0$ then *x'-ct'* = 0 even though *x-ct* != 0.

Einstein:
"An analog reflection of a light beam propagating along the negative *x*-axis gives the condition":

$$(x'+ct') = \mu(x+ct) \qquad (4)$$

Even here one must state that

$$\mu \mathrel{!=} 0 \qquad (4.1)$$

Below, I use A, B instead of a, b:

Einstein:
"If one now adds and subtracts the equations (3) and (4) one obtains:

$$x' = Ax - Bct \qquad (5.1)$$
$$ct' = Act - Bx \qquad (5.2)$$

where for convenience we have introduced

$$A = (\lambda + \mu)/2 \text{ and } B = (\lambda - \mu)/2.$$

Now our task would be solved if we knew constants A and B. We find them through the following considerations: "

Furthermore, Einstein uses the following three conditions:

c1) $x' = 0$
c2) $t = 0$
c3) $t' = 0$

My mathematical proof:
From c1) and (5.1) $\to$ $x = ctB/A$
From c3) and (5.2) $\to$ $x = ctA/B$
$\to$ $B/A = A/B$ $\to$ $A \neq 0$ and $B \neq 0$ and $A^2 = B^2$
$\to$ $((\lambda+\mu)/2)^2 = ((\lambda-\mu)/2)^2$ $\to$ $(\lambda+\mu) = +-(\lambda-\mu)$
$\to$ $\lambda+\mu = \lambda-\mu$ $\to$ $2\mu = 0$ $\to$ $\mu = 0$ contradicts (4.1) or
$\to$ $\lambda+\mu = -\lambda+\mu$ $\to$ $2\lambda = 0$ $\to$ $\lambda = 0$ contradicts (3.1)

This means that the derivation of Lorentz transformations is incorrect!

Derivation of the Lorentz transformations: example 5

Below we follow *[3], page 125; Appendix;*
A simple derivation of the Lorentz transformation

In his presentation of the theory of special relativity, Einstein comes finally to Lorentz transformations:

$x' = (x - vt)\gamma$ (LT$_1$)
$t' = (t - vx/c^2)\gamma$ (LT$_2$)

where $\gamma = 1/(1 - v^2/c^2)^{1/2}$ is called the Lorentz factor, c is the speed of light.
This factor is: $\gamma > 1$ (v > 0), $\gamma < +\infty$ (v < c).

Einstein:
"A light running along the positive x-axis are transmitted from the equation"

$x = ct$ eller $x - ct = 0$ (1)

Similar to the second coordinate system.

$x' = ct'$ eller $x' - ct' = 0$ (2)

Einstein:
"The points in space-time (events) that satisfy (1) must also satisfy (2). This is obviously the case if we generally have the relationship

$$(x'-ct') = \lambda(x-ct) \qquad (3)$$

where λ is a constant. For according to (3) becomes *x'-ct' equal to zero if x-ct is equal to zero.*"

"An analog reflection of a light beam propagating along the negative x-axis gives the condition":

$$(x'+ct') = \mu(x+ct) \qquad (4)$$

Einstein:
"If one now adds and subtracts the equations (3) and (4) one obtains:

$$x' = Ax - Bct \qquad (5.1)$$
$$ct' = Act - Bx \qquad (5.2)$$

where we have introduced for convenience

$$A = (\lambda+\mu)/2 \text{ och } B = (\lambda-\mu)/2.$$

Now our task would be solved if we knew constants A and B. We find through the following considerations: "

Furthermore, Einstein uses the following three conditions:

c1) $x' = 0$
c2) $t = 0$
c3) $t' = 0$

My mathematical proof:
LT_1, c1 → $x' = 0$, $x = vt$
LT_2, c1 → $x' = 0$, $t' = (t-vx/c^2)\gamma$

LT_1, c2 → $t = 0$, $x' = x\gamma$
LT_2, c2 → $t = 0$, $t' = -vx\gamma/c^2$

LT_1, c3 → $t' = 0$, $x' = (x-vt)\gamma$
LT_2, c3 → $t' = 0$, $t = vx/c^2$

All these results will verify LT_1, LT_2 because the conditions c1, c2 and c3 were all used in the derivation of LT_1 and LT_2.

We take $LT1_1$, c1 and LT_2, c3:
$x = vt$ and $t = vx/c^2$ → $x = vvx/c^2$ → $1 = v^2/c^2$ → $v^2 = c^2$
→ $v = +- c$!

The speed v is speed of referens system S' (or S_2, in my figures) is a material object wich **can not** moves with speed of light!

This means that the derivation of Lorentz transformations is incorrect!

Back to Newton

Derivation of the Lorentz transformations: example 6

Below we follow *[3], page 125; Appendix;*
A simple derivation of the Lorentz transformation

In his presentation of the theory of special relativity, Einstein finally to Lorentz transformations:

$x' = (x-vt)\gamma$ (LT_1)
$t' = (t-vx/c^2)\gamma$ (LT_2)

where $\gamma = 1/(1 - v^2/c^2)^{1/2}$ is called the Lorentz factor, c is speed of light.
This factor is: $\gamma > 1$ (v > 0), $\gamma <$ +∞(v < c).

Einstein:
"A light signal which is running along the positive *x*-axis moves corresponding to the equation

$x = ct$ eller $x-ct = 0$ (1)

Similar applies to the second coordinate system.

$x' = ct'$ eller $x'-ct' = 0$ (2)

Einstein:
"The points in space-time (events) that satisfy (1) must also satisfy (2). This is obviously the case if we generally have the relationship

$$(x'-ct') = \lambda(x-ct) \tag{3}$$

where λ is a constant. For according to (3) becomes *x'-ct' equal to zero if x-ct is equal to zero.*"

It should be specified that $\lambda ! = 0$

Einstein:
"An analog reflection of a light beam propagating along the negative x-axis gives the condition":

$$(x'+ct') = \mu(x+ct) \tag{4}$$

It should be specified that $\mu ! = 0$

Einstein:
"If one now adds or subtracts the equations (3) and (4) one obtains:

$$x' = Ax - Bct \tag{5.1}$$
$$ct' = -Bx + Act \tag{5.2}$$

where for convenience we have introduced

$$A = (\lambda+\mu)/2 \text{ and } B = (\lambda-\mu)/2.$$

Now our task would be solved if we knew constants A and B. We find them through the following considerations: "

Furthermore, Einstein uses the following three conditions:

c1) $x' = 0$
c2) $t = 0$
c3) $t' = 0$

My mathematical proof:
5.1, c1 → $0 = Ax - Bct$ → $Ax = Bct$ → $x = (B/A)*ct$
5.2, c1 → $ct' = -Bx + Act$
5.1, c2 → $x' = Ax$
5.2, c2 → $ct' = -Bx$
5.1, c3 → $x' = Ax - Bct$
5.2, c3 → $0 = -Bx + Act$ → $Bx = Act$ → $x = (A/B)*ct$

We have received:
r1) $x = (B/A)*ct$
r2) $ct' = -Bx+Act$
r3) $x' = Ax$
r4) $ct' = -Bx$
r5) $x' = Ax-Bct$
r6) $x = (A/B)*ct$

We combine and get:
r1, r6 → $A/B = B/A$ → $A \ne 0$ och $B \ne 0$
r3, r5 → $Ax = Ax - Bct$ → $-Bct = 0$ → $Bt = 0$ → $t = 0$
r2, r4 → $-Bx = -Bx + Act$ → $Act = 0$ → $At = 0$ → $t = 0$
→ $x = 0, t = 0, x' = 0, t' = 0$

Then we have the trivial case, when two coordinate systems are in the same point! And then we need no transformations to go from one referens system to another, then we need no theory of special relativity!

Derivation of the Lorentz transformations: example 7

Below we follow *[3], page 125; Appendix;*
A simple derivation of the Lorentz transformation

$$x' = Ax - Bct \quad (5.1)$$
$$ct' = -Bx + Act \quad (5.2)$$

c1) $x' = 0$
c2) $t = 0$
c3) $t' = 0$
→
$$x' = (x-vt)\gamma \quad (LT_1)$$
$$t' = (t-vx/c^2)\gamma \quad (LT_2)$$

My mathematical proof:
c2 should not be used when we are at the starting point → $t > 0$

c1 och LT_1 → $x = vt$
c3 och LT_2 → $t = vx/c2$
→
$v = +-c$
no comments

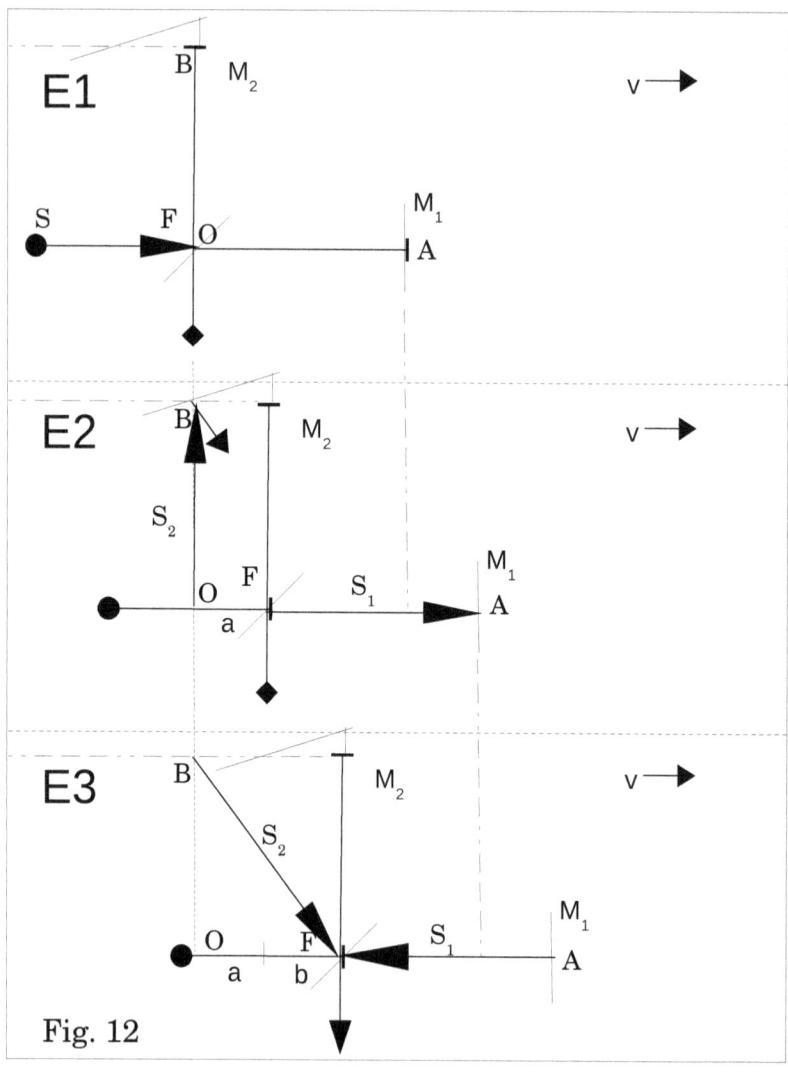

Fig. 12

Michelson-Morley experiment, 1887

E1: Michelson interferometer
Interferometer arms $FA = FB = L$.
A beam of light is sent from S and is splited into two in F. S_1 continues straight on towards A. S_2 goes to B.

E2:
When S_1 reaches A, is reflected on it, and goes back to O. When S_2 reaches B, is reflected on it, and go "back" to F.

Meanwhile S_1 goes against A and reaches this point, the entire system moves with a. Then S_1 traveled distance $L + a$. S_2 traveled distance L.
Note S_1 and S_2 are not reflected at the same time!

E3:
Meanwhile the reflected S_1 goes to F and reach this point, moves the whole system with b. Then S_1 traveled distance $L - b$. S_2 traveled distance
$BF = (OB^2 + OF^2)^{1/2}$.

We now calculate the length of the distances that S_1 and S_2 traveled.

Length(S_1) = $L + a + L - b$.
Length(S_2) = $L + (L^2 + (a+b)^2)^{1/2}$.
$a = Lv/(c-v)$
$b = Lv/(c+v)$
$a+b = 2Lcv/(c^2-v^2)$

Length (S_1) = $2L + a - b = 2Lc^2/(c^2-v^2)$
Length (S_2) = $L + (L^2 + (a+b)^2)^{1/2} = 2Lc^2/(c^2-v^2)$

The length the two light beams pass are the same!

This means that the Michelson interferometer could not detect if there is any ether!
This experiment has been used as an argument for the construction of the theory of special relativity. But as we see now the experiment is based on incorrect assumptions of how the light moves.

There was no sense to build such an interferometer. It was a waste of money! One could have understood from the beginning that the result should be NEGATIV!

We look at the second case, Fig. 13, when the interferometer is rotated by 90 degrees counterclockwise. We do similar calculation of the distances the two light signals pass, like we did in the experiment in Fig. 12.

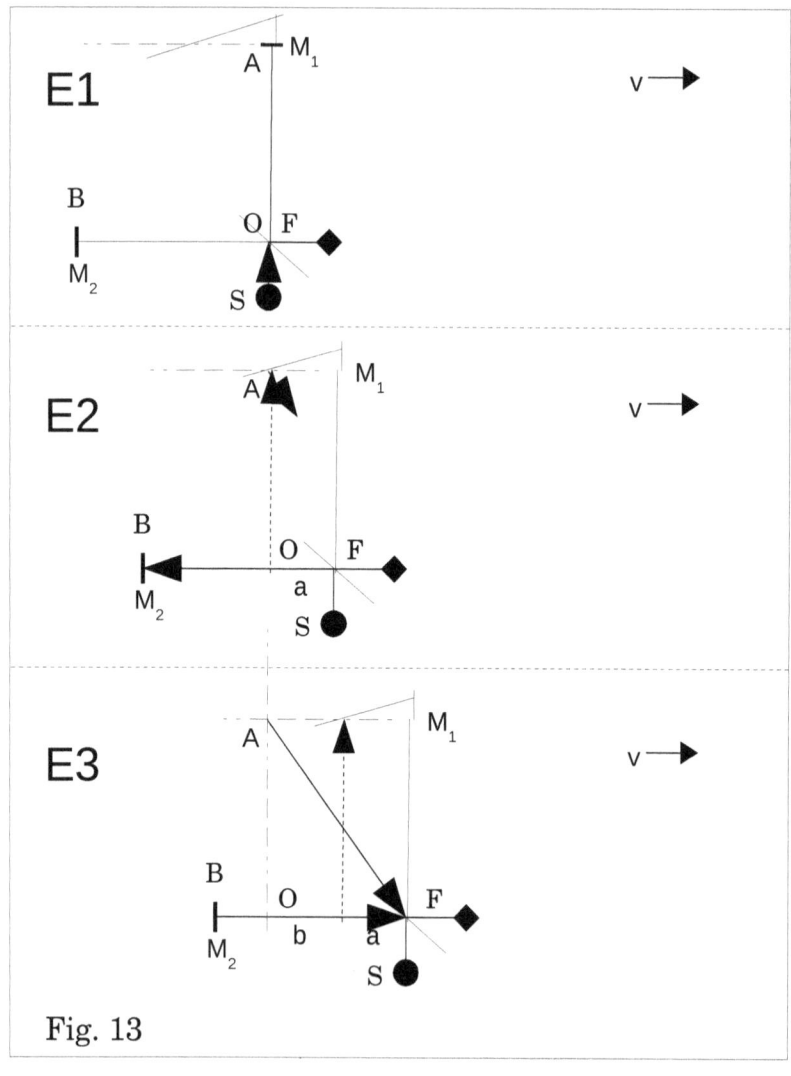

Fig. 13

E1:
Interferometers arms $FA = FB = L$.
A beam of light sent from S is splited into two in F. S_1 continues straight on towards A. S_2 goes to B.

E2:
When S_1 reaches A, is reflected on it, and go "back" toward O. When S_2 reaches B, reflected on, and goes back to F.

Meanwhile S_2 goes towards B and reachs this point, the entire system moves with a. When S_2 traveled distance $L - a$. S_1 traveled distance L.
Note S_1 and S_2 are not reflected at the same time!

E3:
Meanwhile the reflected S_2 goes to F and reach this point, the whole system moves distance b. Then S_2 traveled distance $L + b$. S_1 traveled distance
$AF = (OA^2 + OF^2)^{1/2}$.

We now calculate the length of the distances S_1 and S_2 traveled.

Back to Newton

Length(S_1) = $L + (L^2 + (a+b)^2)^{1/2}$.
Length(S_2) = $L + a + L - b$.
$a = Lv/(c+v)$
$b = Lv/(c-v)$
$a+b = 2Lcv/(c^2-v^2)$

Length (S_1) = $2L + a - b = 2Lc^2/(c^2-v^2)$
Length (S_2) = $L + (L^2 + (a+b)^2)^{1/2} = 2Lc^2/(c^2-v^2)$

The length the two light beams pass are the same!

No matter how you turn the interferometer, the two scattered light signals S_1 and S_2 travel the same distance.
This means that there can be no interference pattern when those two light signals reunited.

It's so obvious why the Michelson-Morley experiment of 1887 has given so-called negative results.

I do not understand how this could happen! After this experiment, it was said that the light-bearing ether does not exist, Einstein used this experiment to develop the special relativity and Albert A. Michelson received the Nobel Prize in Physics in 1907!

Epilogue

In this work, we have analyzed the following:

1) Experiments with two reference systems, S_1 and S_2, standing against each other, and an object in S_1 in which events occur.

2) Experiments with two reference systems, S_1 and S_2, which moves at a constant speed $v > 0$ to each other, and an object in S_1 in which events occur.

We show that there is the same coordinate transformation either the two reference systems, S_1 and S_2, are standing against each other or if they move with constant velocity $v > 0$ towards each other.

3) Time dilation in *[1], [6] and [8]*.

We show how completely wrong one reason about the propagation of light and that the 'time clock' works the same way in the two reference systems, S_1 and S_2, which moves with constant velocity $v > 0$ towards each other. Thus, we show that there is no time dilation in any of the two referens systems!

4) Derivation of the Lorentz transformations in *[7]*.

We show here that the calculations are incomplete and that we come to a contradiction with the theory's assumptions which makes that this derivation is wrong!

5) Derivation of the Lorentz transformations in *[3]*.

We show how this derivation is based on erroneous mathematical assumptions and we come to a contradiction with the theory's assumptions that allows that even this derivation is wrong!

6) The Michelson-Morley experiment, 1887

We show that the experiment was doomed to failure from the beginning, the Michelson interferometer could not detect the Earth's motion around the Sun!

This analysis of the theory of special relativity detects too many errors, too many misinterpretations!

From this, we should conclude that the theory of special relativity is incorrect from the ground, in its entirety!

I am grateful if readers will comment my book.

e-mail address: jan.slowak@gmail.com

Enter the subject: Back to Newton.

Einstein's Theory of Special Relativity - a mathematical impossibility!

Jan Slowak
Jönköping, Sweden
jan.slowak@gmail.com

February 22, 2017

Abstract

Einstein's Theory of Special Relativity is a generally accepted theory that analyses relationships between two inertial reference systems moving at a constant speed against each other. In this paper three special cases are used and inserted into the Lorentz transformations resulting in a mathematical contradiction. This leads to the conclusion that the Theory of Special Relativity is incorrect.

When we study physical phenomena, we always make a mathematical model of them. In such a model there are built-in physical laws that are held together by mathematical tools. If the description of the physical phenomenon is correct then the mathematical model is also correct!

The Theory of Special Relativity deals with the relationship between two inertial reference systems, S and S', moving towards each other at a constant speed $v > 0$.

Every event in these reference systems is determined by four coordinates, three for space and one for time. To determine an event's coordinates in one of the reference systems with the help of event's coordinates in the second reference system one uses Lorentz' transformations:

$$E = (t, x, y, z), \text{ an event in S}$$
$$E' = (t', x', y', z'), \text{ an event in S'}$$

To facilitate understanding of the calculations one usually puts

$$y = y'$$

$$z = z'$$

The Lorentz transformations become:

$$x' = (x - vt)\gamma \qquad (1)$$

$$t' = (t - \frac{vx}{c^2})\gamma \qquad (2)$$

where $\gamma = \frac{1}{\sqrt{1-\frac{v^2}{c^2}}}$ is called Lorentz factor.

The derivation of the Lorentz transformations is done with the assumption that these transformations must be linear:

$x' = Ax + Bt$
$t' = Cx + Dt$, where A, B, C and D are constants.

To solve this system of equations one uses three special cases:

$$x' = 0, x = vt \qquad (3)$$
$$x' = -vt', x = 0 \qquad (4)$$
$$x' = ct', x = ct \qquad (5)$$

where c is the speed of light.
This derivation is in [1]

But if Lorentz transformations were calculated using 3, 4 and 5 then these three special cases should satisfy Lorentz transformations 1 and 2 without leading to a mathematical contradiction.

My proof:
From 3 and 1 $\Rightarrow 0 = (vt - vt)\gamma \Rightarrow 0 = 0$ OK
From 3 and 2 $\Rightarrow t' = (t - \frac{v(vt)}{c^2})\gamma \Rightarrow t' = t(1 - \frac{v^2}{c^2})\gamma$

From 4 and 1 $\Rightarrow -vt' = (0 - vt)\gamma \Rightarrow -vt' = -vt\gamma \Rightarrow t' = t\gamma$

From 4 and 2 $\Rightarrow t' = (t - \frac{v \cdot 0}{c^2})\gamma \Rightarrow t' = t\gamma$

But the result from 3 and 2 is $t' = t(1 - \frac{v^2}{c^2})\gamma$ and the result from 4 and 2 is $t' = t\gamma$

$\Rightarrow 1 - \frac{v^2}{c^2} = 1 \Rightarrow v = 0$

This result, v = 0, is in contradiction with the theory's assumption that the two reference systems move towards each other with constant velocity v > 0!

This shows that the derivation of the Lorentz transformations is wrong and with it even the Theory of Special Relativity.

REFERENCES

[1] Modern Physics, second edition, Randy Harris, 2008, page 14-15.

www.ingramcontent.com/pod-product-compliance
Lightning Source LLC
Chambersburg PA
CBHW050237230526
45470CB00005B/2003